CW00829080

TRANSPORTER BRIDGES

By

HENRY GRATTAN TYRRELL, C.E.

Bridge and Structural Engineer

EVANSTON, ILL.

Author of Mill Building Construction, 1900 ; Concrete Bridges and Culverts. History of Bridge Engineering. Mill Buildings. Artistic Bridge Design, etc., etc.

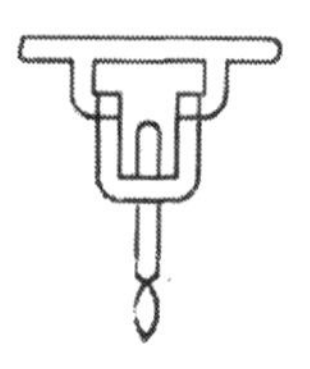

Published by
The University of Toronto Engineering Society
Toronto: 1912

BOOKS BY THE SAME AUTHOR

History of Bridge Engineering.
Cloth Binding 6 by 9 inches
500 Pages 330 Illustrations. Indexed. Published by the Author. Price, $4 00, Postpaid

Mill Building Construction (1900)
Cloth Binding 6 by 9 inches Price $1 00
(Out of Print)

Concrete Bridges and Culverts.
Flexible Leather 4½ by 6¾ inches.
272 Pages 66 Illustrations Price $3 00

Mill Buildings. A treatise on the Design and Construction of Mill Buildings and Other Industrial Plants
Cloth Binding 6 by 9 inches.
490 Pages 652 Illustrations Price, $4 00

Artistic Bridge Design A Systematic Treatise on the Design of Modern Bridges, according to Aesthetic Principles
Cloth Binding 6 by 9 inches
300 Pages 250 Illustrations Price, $3 00

Engineering of Shops and Factories.
Cloth Binding 6 by 9 inches
400 Pages 150 Illustrations
(In Press)

PAMPHLETS

Vertical Lift Bridges.
Paper Covers 6 by 9 inches
16 Pages Illustrated Price, 50 cents

The Elizabethtown Bridge. The Longest Simple-Truss Span (1909)
Paper Covers 6 by 9 inches.
24 Pages. Five Pages of Illustrations
Price, 50 cents

Transporter Bridges

Transporter bridges of one kind or another have been in use for many centuries. The earliest ones are probably those of India, called by the natives "tarabita," and used by them for crossing streams or mountain gorges. They consisted of a single rope made of hides or fibres fastened at the ends to trees on shore, and from this rope a basket was suspended and drawn back and forth by a smaller cord. Similar bridges are known to have been used many centuries ago by the natives of Peru and adjoining countries in South America.

A contrivance of this kind, though with more detail, was used by Faustus Verautius about 1620. Wooden posts or towers were planted at each side of the stream, and between the tops of the posts was stretched a rope or cable from which a basket or car was suspended by two trolleys, one attached to each end of the conveyor. Landing platforms or brackets were fastened to the towers at the proper elevation to correspond with the floor of the moving car, and these platforms were reached by wooden ladders from the ground. A smaller endless hauling rope passing over a pulley on the top of each tower, hung loosely in the car, and by means of this rope, the car and its load was drawn back and forth by one of the occupants. These early inventions are the prototypes of the more elaborate modern cableways and transporter or ferry bridges.

The principles contained in the primitive bridges described above, were revived in England during the first part of the nineteenth century, when patents were granted to Smart, Fisher & Leach for three different types of railroad drawbridges. Drawings and models were made in 1822 for Smart's bridge, and Fisher's patent for an aerial ferry was taken out two years later, but a more elaborate drawing, though somewhat similar to Smart's, was that proposed by Harvey Leach for a suspension railway ferry. His plan showed a series of spans 300 to 400 feet in length, above which cables were suspended which supported a horizontal track or runway high enough above water to leave the desired head room below, for ships and river craft. From the upper runway, a platform the full length of one span was suspended between the piers, and this platform, with its load, was capable of being moved back and forth as desired. A patent for an aerial railway bridge to cross the East River at New York was granted about 1852 to H. N. Houghton, of Bergen, N. J., who proposed placing a number of heavy stone piers in the river, with truss spans thereon, and a clearance under the spans of 150 to 200 feet for ships. Instead of expensive approaches to a high level bridge, he proposed suspending a moving platform for a double line of railway, and making this platform long enough to carry whole trains of cars. The obstruction which this plan offered to shipping was from the river piers only, the space between them being always open except for the occasional passing of the moving platform. To avoid all river obstruction, even that offered by the piers, it was proposed by Morse in 1869 to cross the East River with a single suspension span having a clear opening of 1,410 feet and an under height of 140 feet, travel being conveyed, not over the bridge, but on a moving platform suspended from the upper runway deck. In other respects his design was similar to Houghton's, for it showed no inclined approaches, but merely a suspended plat-

Fig. 1.

form about 150 feet long to travel back and forth at intervals between New York and Brooklyn. As far as the length of span was concerned, the design was not remarkable, for suspension bridges with lengths of 1,200 feet or more between the towers, had previously been built, and several had been proposed of much greater length, including the notable design of M. Oudry, for crossing Messina Straits with four suspension spans of 1,000 meters each.

In 1873, Mr. Charles Smith, manager of the Hartlepool Iron Works, of Hartlepool, England, designed a transporter bridge with cantilever trusses to cross the Tees at Middlesborough, with a center span of 650 feet and a total length of 1,000 feet. His plans were endorsed by no less an authority than Benjamin Baker, but because of insufficient funds, the project was not carried to completion and a steam ferry was installed instead, at a cost of less than $50,000 On account of the publicity given to this project, it has often, but incorrectly, been referred to as the first design for a transporter bridge

Five years later, an elaborate plan for a transporter bridge over the Thames, was prepared by L Mills and A Twyman of North Shields, with a center opening 200 feet in width and 80 feet high The upper platform, reached by elevators in the towers, was to have provision for pedestrian travel, so that foot passengers could cross at all times

As transporter bridges are especially suitable for crossing harbor entrances at the sea coast, the type had for many years been advocated for the water courses at New York, and in 1885, Mr John F Anderson published a design, Fig 1, for crossing the Hudson by means of a moving platform suspended from a high level track, supported on pairs of cylinder piers The platform was to be long enough to always be in contact with three sets of piers, thereby insuring lateral stability In other respects the design was quite similar to those previously prepared by Harvey Leach and H N Houghton and to Haege's plan for a rolling railway bridge • Two years previous to this (1883), Mr Gustav Lindenthal had been granted an American patent on a transporter bridge with a traveling suspended car

During the year 1894, two important passenger cableways were erected, one near Knoxville, Tennessee, and the other at Brighton Dyke, England, the car on the former one moving on a cable with steep incline The cableway crossing Devil's Dyke at Brighton, designed by W J Brewer, had a clear center span of 650 feet, the type being selected because conditions would not permit the expense of a regular bridge An upper unstiffened cable over the towers, with a sag of only 26 feet, supports all the load, and two lower horizontal cables suspended therefrom by one-inch steel bars, carry the trolley at a height of 230 feet above the valley at the deepest part The car is only 5 by 7 feet, to hold from eight to twelve passengers, and it is hauled back and forth by a smaller rope, making the passage in 2½ minutes After its completion, 720 people were taken across and back in 2½ hours.

The development and introduction of transporter bridges is due chiefly to the enterprise of Ferdinand J Arnodin, proprietor of the iron works at Chateauneuf, France, who during the past twenty years has erected at least eight of these structures at Bilbao, Bizerta, Rouen, Rochefort, Nantes, Marseilles, Newport, and Tangier The first of these, between Portugalete and Los Arenas, over the mouth of the Nervion or Bilbao River, on the coast of the Bay of Biscay, about ten miles from Bilbao, Spain, was completed in 1893 (Fig 2) The metal towers rising at each side of the river are 525 feet apart on center, and a horizontal runway 131 feet above water is supported by cables passing over the towers and anchored to blocks of masonry. The horizontal runway, in addition to hanging from the cables above it, is supported at each end for about one-quarter the span length, by stay cables from the towers. The total moving dead load is 40 tons, and the car which carries 150 passengers, crosses from one side to the other in one minute. The design is the combined work of Arnodin and Palacio.

The second of M. Arnodin's designs crosses a canal at Bizerta in Tunis, and replaced the ferry which was guided by a cable during transit. The bridge was commenced in 1896 and completed two years later. It is similar in outline to that at Bilbao but with a shorter span, the distance between the towers, which are 213 feet high, being only 355 feet, though the under clearance of 148

Fig. 2.

feet is slightly greater than the previous one. The car is 32 feet by 24 feet, and it is moved by a steel cable and steam power. The itemized cost was as follows:—

Steel, without machinery	$95,500
Machinery	8,600
Miscellaneous work	2,600
Duty	5,000
	$111,700

It was severely tested by a cyclone in 1898, but remained uninjured. It was proposed by the government in 1904 to take the structure down and to re-erect it at Bordeaux or Brest, changing the motive power from steam to electricity.

The third of M. Arnodin's designs was completed a year later (1899) over the Seine at Rouen, being quite similar to his first one at Bilbao, though with a larger capacity and cost. The clear distance between docks is 436 feet, and between tower centers 469 feet, while the under clearance above the dock is 164 feet. The towers, which are 221 feet high, support twelve steel wire cables from which the horizontal runway is suspended. The moving platform is 33 feet long and 42 feet wide, and has a weight of 37 tons when empty, and 45 tons when loaded. It has a capacity for 200 persons and 6 vehicles, and is suspended from the trolley by thirty cables. The car, Fig. 3, can be made to cross the channel in 45 seconds, though the usual time is about 80 seconds. Its maximum daily

Fig. 3

service is 240 trips to and fro, carrying 300 vehicles and 10,000 passengers. It cost $180,000 and the schedule of tolls thereon is as follows:—

First-class passengers	2 cents
Second-class passengers	1 cent
Two wheeled rig	6 cents
Four wheeled rig	8 cents
One-horse cart, empty	5 cents
One-horse cart, loaded	8 cents
Two-horse cart, empty	7 cents
Four-horse cart, loaded	13 cents

During the following year (1900), M. Arnodin proposed several transporter bridges in England, one over the Ribble Navigation, and another over the Tyne between North and South Shields, with a span of 650 feet, having the co-operation of Mr. C. H. Gadsby on the latter one. In June, 1901, M. Arnodin took out American patents on a transporter bridge of cantilever type with suspended center span, similar to that which he completed in 1903 over the Loire River at Nantes, which was the first of its kind to be completed. The platform of the bridge at Nantes is supported at intervals of 15 feet by stay cables from the tower tops, and the projecting cantilever arms are connected by a suspended span 113½ feet long. The front and rear arms of the cantilever are 175 and 82 feet long respectively, and the rear end is tied down with wire ropes to the anchor masonry. The distance between the tower centers is 462 feet and the total length 626 feet, the clear height underneath for ships being 165 feet. The trusses are 26 feet apart and the car suspended from them is 40 by 40 feet, with a maximum capacity of 60 tons. It is operated by an electric motor on the truck, and will cross the water in one minute. It cost $199,000, and the schedule of charges is as follows:—

Pedestrians	1 cent each
One-horse cart, empty	5 cents each

Fig. 4

Two-horse cart, empty	7 cents each
One-horse cart, loaded	8 cents each
Two-horse cart, loaded	10 cents each
Wagons, loaded	12 cents each

The most daring project for a transporter bridge ever undertaken was that which appeared in 1903 for crossing the Gironde River at Bordeaux, with a single arch of 1,412 feet (Fig. 4) the span being about the same as that designed by Morse in 1869 for crossing the East River at New York. The proposed Bordeaux bridge consisted of a pair of metal arches, in vertical planes and about 80 feet apart, from which the runway deck was suspended, leaving a clearance of 150 feet beneath it. The total rise of the arch was 328 feet (100 meters) and that part of the ribs above the runway were lune-shaped with three hinges, the longitudinal distance between the end pins being shortened by this arrangement to 990 feet, similar to that used about the same time for the Austerlitz arch bridge at Paris. The clear distance between docks was to be 1,312 feet, and that between towers centres 100 feet additional, making it longer than any arch yet built. The runaway deck had provision for a footwalk but was without stiffening trusses, and it supported a double line of track, so that cars might start from each side of the river at the same time Towers were 33 by 112 feet, and 164 feet high and they had elevators to carry passengers to the upper crossing A somewhat similar bridge, though not a transporter, was proposed by Max Ende for crossing the Thames at London In the latter case, instead of using suspended cars, travel of all kinds was to be raised and lowered on elevators running on inclined tracks at the ends, and descending into pits below the streets at each side of the river (See Tyrrell's *History of Bridge Engineering*, page 336)

The Marseilles transporter of 1904 is similar to that at Nantes, with cantilever arms and a centre span Towers stand on cylinders and are 541 feet apart on centre, and the runway deck which was erected by cantilever method, is 160 feet above the water

Up to this time, transporter bridges had not been used in America though they are quite as suitable for harbor entrances here, as in Europe, but in 1905, the first and only one on this side of the Atlantic, was completed, over the ship canal from Lake Avenue, Duluth, to Minnesota Point The site had been a perplexing one for bridge engineers, for they had wrestled with the problem for fifteen years or more In 1890, Mr A P Boller made plans for a swing bridge revolving horizontally on a shore pier, with a clear opening of 200 feet and deck 20 feet above water, the estimated cost being $400,000 As this was more than the city cared to spend, a prize of $1,000 was offered for the best design for a movable bridge to suit the place, and in response, twenty or more plans were prepared and submitted Estimates were also submitted for a double tunnel, varying in amount from $500,000 to $1,300,000 In 1899, when the bridge at Rouen had been completed and publicly illustrated, the city engineer, Mr T F McGilvray, prepared drawings for a similar bridge to cross the channel at Duluth with a clear opening of 300 feet, or 383 feet between tower centres French engineers having discovered the lack of rigidity in suspended tracks for short span transporter bridges, were planning a new one for Nantes on the cantilever principle with an intermediate truss, similar in some respects to that proposed by Charles Smith in 1873 for Middlesborough An alternate plan to that prepared by the Duluth city engineer, was also made in 1901 by a local agent of the American Bridge Company that was tendering for a construction contract In this plan (Fig 5) rigid framing was used throughout, the distance between front tower legs being 393 feet, 9 inches, and the height beneath the bridge 135 feet, as at New York City As the structure from its exposed position would certainly be subject to severe gales, every effort was made to secure rigidity, double riveted web systems being used in the trusses and stiff braced members for the car suspenders, thus preventing it from swaying in the wind Several types of construction had previously been used for transporter bridges in Europe those at Bilbao, Bizerta, and Rouen, being suspensions, Nantes and Marseilles, cantilevers, and the proposed one at Bordeaux an arch; and it is interesting to note that still another type—a simple truss—was selected for the bridge at Duluth, to which, with its com-

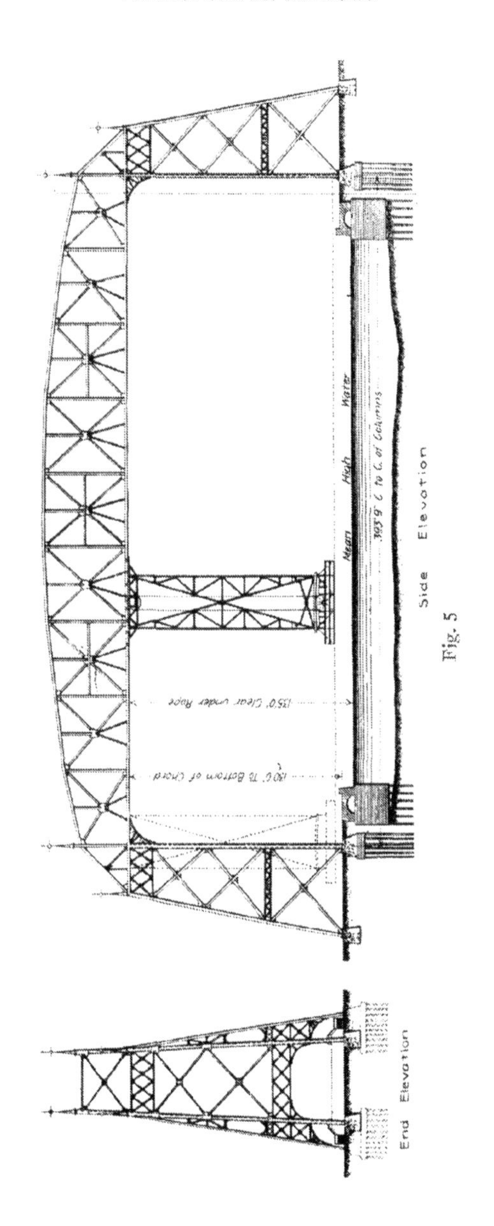

Fig. 5

paratively short span, it is well adapted, at the same time making a patent more easily obtainable. In several European designs, the moving car passes through the towers which are braced laterally to resist wind pressure, but in the Duluth bridge the platform runs in between the front columns only, without interfering with the tower bracing on the outer side. The car is suspended from a rigid trolley frame, the wheels of which, mounted on roller bearings, run on rails inside the box chords. The car is propelled by electric power from two different sources, a one-inch steel rope being fastened to each tower and wound on a drum attached to the moving part. When moved by electric power, the car crosses the canal in one minute, but it also has hand power for emergency. Both trolley and car run against air buffers at each end, and jar is further avoided by links in the suspenders near the deck. Construction was under way more or less for four years and after much delay, change of contractors, and revision of plans, it was finally completed in 1905 at a cost of $100,000, though a number of ornamental features that were at first intended were omitted.

Another very interesting design for a modified type of transporter bridge appeared in 1905, the invention of Abraham

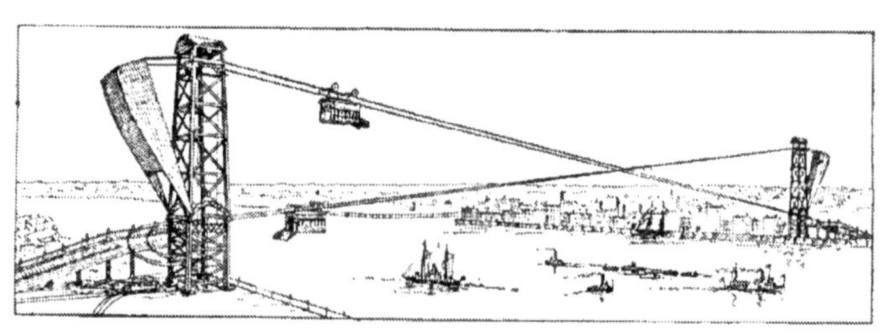

Fig. 6

Abelson, of New York, the essential principle of which was gravity car motion. As will be seen from the illustration (Fig. 6) at each side of the river or ravine, towers are erected, to the base of which double cables are attached, which, after crossing to the opposite side, pass over saddles at the tower tops and fasten to counterweights hinged to the opposite side. This method of fastening the cables avoids the obstruction caused by carrying them back in the usual way to anchor blocks on shore, and at the same time holds the cables taut though not absolutely rigid. The same anchorage principle is now applied in modern freight cableways and is effective in allowing for expansion or slight variation in the cable length. The effect of the hinged weights counteracting the pull on the cables is to produce vertical reactions on the towers, without any tendency to tipping. An elevator operates in each tower, and a car on crossing from the other side makes a detour around the tower to the land side and is then loaded on the elevator by which it is lifted to the upper level The car trucks are swiveled that they may be easily removed from contact with the cables and attached to them again in their raised position The cars are not connected with each other and their speed may be regulated by the operator

Although transporter bridges have been projected in Great Britain for forty years or more, none were built there prior to 1905, but since that time four fine structures have been completed at Runcorn, Newport, Warrington and Middlesborough, three of which were the designs of English engineers A suspension over the Mersey at Runcorn was proposed in 1817 by Thomas Telford though never built His design showed a clear span of 1,000 feet with stone towers, a roadway 30 feet wide, and a clearance of 70 feet above the water, the estimated cost being $450,000 But the river at that place remained unbridged until 1868, when the London and North Western Railway erected a bridge 1,300 feet long in three spans, with an under clearance of 78 feet, to the deck of which pedestrians had access by stairs at the ends The new suspension transporter crosses the Mersey and the Manchester Ship Canal between Widnes and Runcorn with a span of 1,000 feet, and an under clearance of 82 feet, or slightly more than that at the railway bridge near by, and it is now the longest highway span in Great Britain It consists of a stiffened track, hung from cables passing over metal towers on shore, high enough to permit ships to pass under, the cables being anchored back into blocks of masonry The two main cables are 12 inches in diameter and are cradled according to the method first used by John A Roebling in 1844 Each cable is composed of nineteen smaller ones, and the whole is wrapped and protected from the weather by canvas and bitumen The angle of inclination which the cable makes with the vertical is different at each side of the tower, and the stress in the back stays is, therefore, about 12 per cent greater than in the span Where they bear on the saddles at the tower tops, the cables and saddles are clamped together to prevent slipping and to insure vertical reactions The cables at the ends are attached to bars or links which are anchored into blocks of masonry, the bars being embedded solid in concrete after receiving the stress from dead load The double towers on each shore are gracefully proportioned tapering out towards the base like trunks of great trees, and they are connected by ornamental portals and diagonal bracing Each one is composed of four columns forming a rectangle 30 feet square at the base, the transverse distance between tower centres being 70 feet A stair at each end gives access for pedestrians to the elevated foot walk Under each tower leg is a cast iron cylinder 9 feet in diameter with 1 1-8 inch metal, the interior of the cylinders being filled solid with concrete The stiffening girders are 18 feet deep and 35 feet apart, the two halves being connected at the centre by hinges The moving platform or car is 24 feet wide and 55 feet long, with a single road and a covered space for pedestrians at one side. It weighs 120 tons and the usual time for crossing is 1¾ minutes, the moving being controlled by a man in an elevated cabin on the car above the roadway. The car is suspended from a trolley 77 feet long, mounted on wheels 18 inches diameter, and running on the track of the elevated deck. It is propelled by an electric motor on the trolley which is supplied with power from a generating station near by. The engineers were J. J. Webster and J. T. Wood, and the contractor, Sir William Arrol & Co., the total cost including approaches being $665,000. It was formally opened in May, 1905.

A bridge over the Usk at Newport in South Wales, designed by F. J. Arnodin and R. H. Haynes was under construction at the same time as the last one described, but though sanctioned by Parliament in 1900, work was not completed until 1906. A stone bridge in five spans was placed over the river in 1800, and it was repaired and widened in 1866 and 1882, and yet another bridge or better crossing facilities were needed. Preliminary designs and estimates for bridges of different sorts showed that a high level structure would cost $6,250,000, and a low level bridge with a swing span, about $3,500,000, the cost in both cases being much greater than the authorities cared to pay. The design for a transporter bridge which was accepted is of the suspension type, with a length of 645 feet between tower centres, and 592 feet in the clear, the distance between centre of anchorages being 1,545 feet. The headroom above the water is 177 feet, and the design in general is somewhat similar to Arnodin's other suspensions. Towers are pin-ended and their tops are 242 feet above the approaches and 269 feet above low water. Each tower contains 277 tons of steel, and stands on four cylinder piers. Sixteen smaller cables were used instead of a single large one at each side, as on the bridge at Runcorn, conforming with the usual French custom. The traveler, which is mounted on sixty cast steel wheels, is 104 feet long and 26 feet wide, and from it is suspended the moving car, 33 feet long and 40 feet wide, weighing 51 tons. It is capable of carrying a live load of 66 tons and its maximum rate of travel is 10 feet per second, being propelled by wire rope and electric power. The cost in detail was as follows:—

Foundations	$95,600
Shore abutments	18,400
Superstructure	140,000
	$254,000

The third bridge of the kind in England crosses the Mersey at Warrington, 18 miles from the larger one at Runcorn. It was erected chiefly for the convenience of employees at the manufactury of Joseph Crosfield & Son on a strip of land known as Tongueland, between adjoining bends in the river. It is of suspension type and has a span between towers of only 250 feet, with an under clearance of 75 feet as required by government. Stiffening trusses are without hinges and were proportioned according to the theory of Merriman and Jacoby. Single cables 7 inches in diameter of plough steel wire were used at each side, the economical sag for the specified load being found by trial to be one-twelfth of the span The horizontal track is suspended from the cables by 1¼-inch rods, 10 feet apart, and from this track the car, which has a capacity of only 5,000 pounds, is hung The accepted design was submitted by Thomas Piggott & Co , of Birmingham, the resident engineer on the work being James Newall

The transporter bridge over the Tees between Middlesborough and Port Clarence, forming a connection between North Yorkshire and Durham County, though proposed in 1873, and elaborate plans then prepared, was the last of four in England to be erected, for it was not formally opened until October, 1911, after twenty-seven months of actual construction It consists of double cantilevers on metal towers anchored at the rear ends to blocks of masonry, and in many respects is similar to the design previously made for the same site by Charles Smith of Hartlepool The span between centre of towers which are 225 feet high, is 570 feet, and the total length is 850 feet An elevated foot walk over the bridge is reached by stairs in the towers The moving car is 41 by 39 feet, with space for six carriages and 600 persons, and it is moved by an endless rope It contains 2,600 tons of steel with 600 tons in the foundation, and the total cost was $408,000 It was designed by the Cleveland Bridge and Engineering Co , and built by Sir William Arrol & Co Another bridge of suspension type with pin-ended towers 390 feet apart on centres, was erected at the Kiel Dockyard in 1911

From the above brief descriptions, it appears that transporter or ferry bridges have a definite use, and are especially applicable for harbor entrances on the sea coast or at other exposed positions, where ships which are unfamiliar with local drawbridge signals, are frequently entering, or where they may be driven for safety during storms To shipping, they offer nearly all the advantages of a high level fixed bridge, and still permit land travel to cross at about water level, at the same time saving the expense of inclined approaches to a high level structure

Illustrations are from Tyrrell s *History of Bridge Engineering Engineering News* and *The Scientific American*

www.ingramcontent.com/pod-product-compliance
Ingram Content Group UK Ltd.
Pitfield, Milton Keynes, MK11 3LW, UK
UKHW021400070125
3997UKWH00007B/98